THE LIBRARY OF THE PLANETS™

URANUS

Luke Thompson

The Rosen Publishing Group's
PowerKids Press™
New York

Published in 2001 by The Rosen Publishing Group, Inc.
29 East 21st Street, New York, NY 10010

First Edition

Book Design: Michael Caroleo and Michael de Guzman

Photo Credits: pp. 1, 4 PhotoDisc; p. 7 Davis Meltzer/NGS Image Collection; p. 8 Courtesy of NASA/JPL/California Institute of Technology; p. 11 (Roman god Uranus) Michael R. Whalen/NGS Image Collection; p. 11 (Uranus) Davis Meltzer/NGS Image Collection; p. 12 (Sun) PhotoDisc, p. 12 Courtesy of NASA/JPL/California Institute of Technology (digital illustration by Michael de Guzman); p. 15 Davis Meltzer/NGS Image Collection; pp. 16, 19, 22 Courtesy of NASA/JPL/California Institute of Technology; p. 20 (rings) © CORBIS, p. 20 (Uranus with ring) Courtesy of NASA/JPL/California Institute of Technology (digital illustration by Michael de Guzman).

Manufactured in the United States of America

Contents

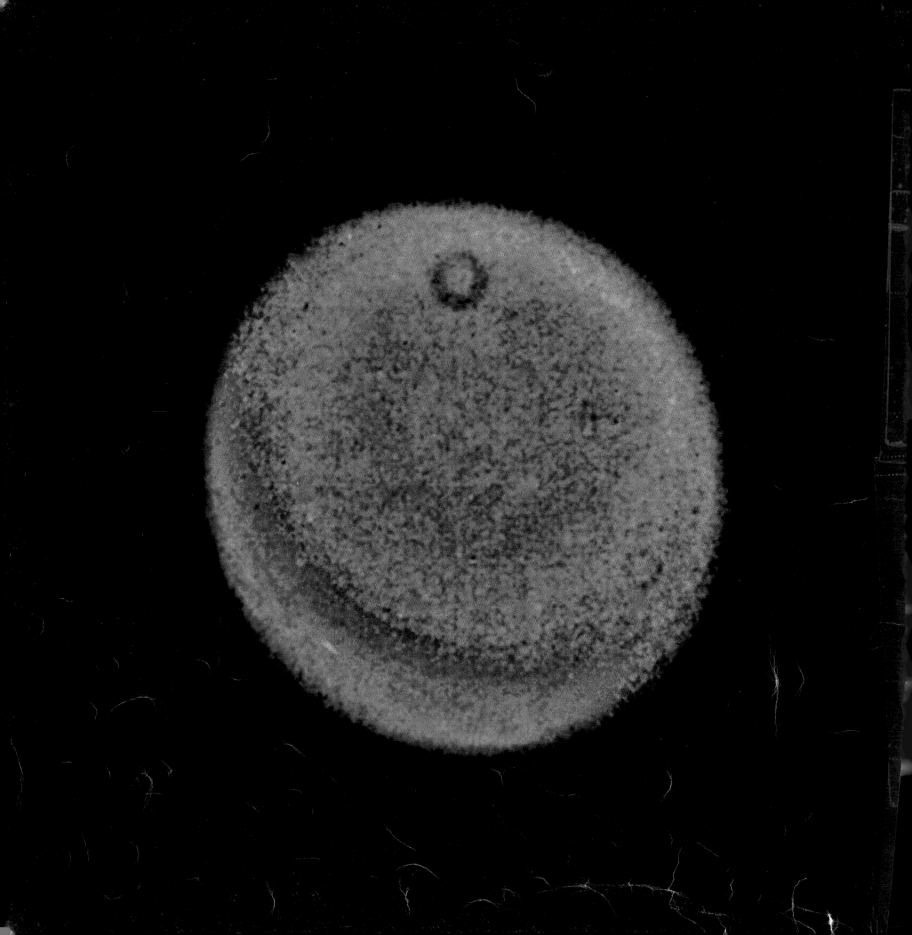

A Small, Blue Dot

Uranus is one of the nine planets in our **solar system**. The nine planets in our solar system circle around the Sun. The Sun is a star that gives off light and heat. Uranus is the seventh planet from the Sun. It is known as one of the outer planets. The other outer planets are Jupiter, Saturn, Neptune, and Pluto.

Uranus is over one and a half billion miles (2.4 billion km) away from Earth. Uranus is too far away from Earth to be seen without a **telescope**. Through a telescope, Uranus looks like a small, blue dot. Pictures of Uranus show that it is a pale, blue ball. Uranus has a bright system of clouds. Its clouds are the brightest in the outer solar system. The clouds are made of a **gas** called **methane**.

Uranus is about 1.8 billion miles (2.9 billion km) from the Sun.

One of the Giants

Uranus is the third largest planet in the solar system. It is one of the four gas giants. The gas giants are large planets that are made up mostly of gas. These planets are Jupiter, Saturn, Neptune, and Uranus. Uranus is also much bigger than Earth. If Uranus were hollow, 50 Earths could fit inside it. If you were to walk in a straight line around the planet Uranus, you would have to walk 99,787 miles (160,592 km). That is four times as far as you would have to walk to travel around Earth. Even though it is a big planet, Uranus is not very heavy. This is because it is made mostly of gas.

Uranus is much bigger than Earth. Uranus is 31,770 miles (51,129 km) across, while Earth is 7,928 miles (12,759 km) across.

URANUS

EARTH

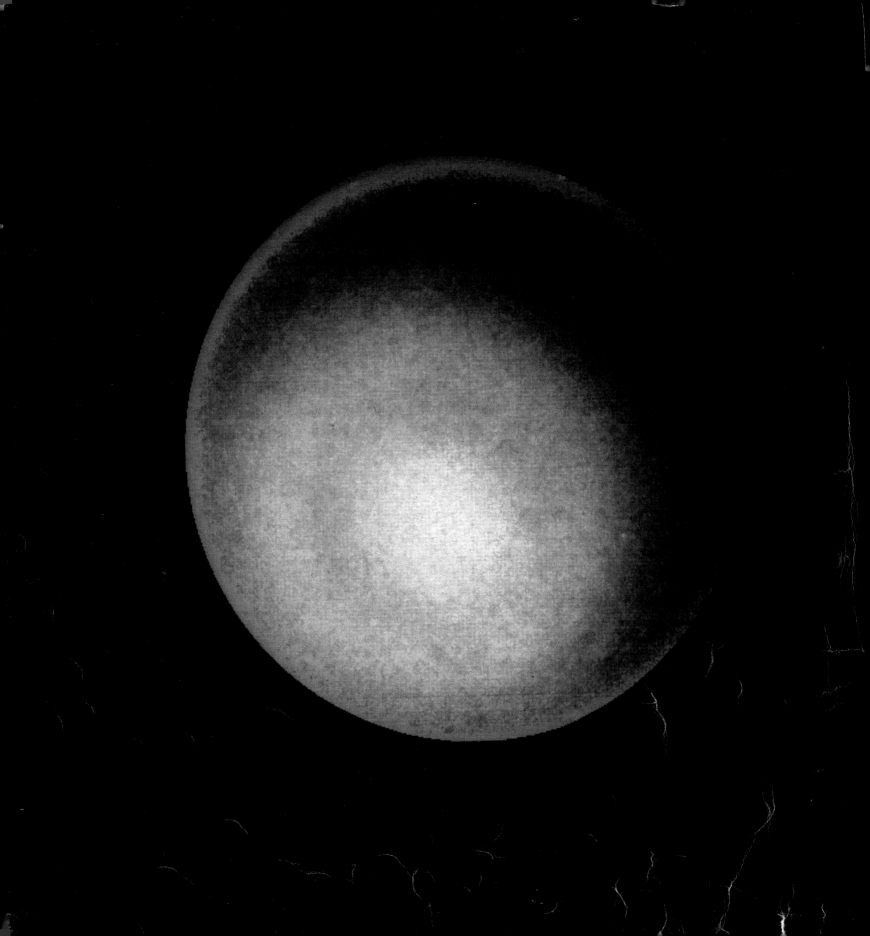

Cold and Dark

Uranus is one and three-quarter billion miles (2.8 billion km) from the Sun. It does not get much sunlight or heat. Light and heat are forms of energy. Here on Earth, we get energy from the Sun. Plants and animals need the Sun's energy to live. Uranus is too dark and too cold for these things to live on it. Temperatures on Uranus can get as cold as -323 degrees Fahrenheit (-197 degrees C). The coldest temperature recorded on Earth is only -100 degrees Fahrenheit (-73.3 degrees C). From Uranus, the Sun looks much smaller and darker than it does from Earth.

This image of Uranus was taken by the Voyager 2 space probe. A space probe is a small spacecraft that travels in space and is steered by scientists on the ground.

Discovering Uranus

Uranus was discovered by William Herschel in 1781. William Herschel was an **astronomer**. An astronomer is a scientist who studies the night sky, the planets, moons, stars, and other objects found there. When Herschel looked through his telescope at the stars, he noticed that one of the stars looked like it was moving faster than the others. It turned out that he wasn't looking at a star. He was looking at a planet.

Uranus didn't have a name for a few years after it was discovered. The other six planets had been named after ancient Greek or Roman gods. The seventh planet was finally named after the Greek god Uranus. Uranus was the god of the sky and the heavens. He was also the father of Saturn. Saturn is the name of one of the other gas giants.

Uranus is named after the Greek god of the sky and the heavens. Behind him is a drawing of the planet Uranus and its ring system.

The Sun

Uranus

Planet	Orbit Time Around the Sun
Mercury	88 Earth days
Venus	225 Earth days
Earth	1 year (365 days)
Mars	1 year and 322 Earth days
Jupiter	12 Earth years
Saturn	29 Earth years
Uranus	84 Earth years
Neptune	165 Earth years
Pluto	249 Earth years

If you weigh 100 lbs. (45.4 kg) on Earth, you would weigh 91 lbs. (41.3 kg) on Uranus.

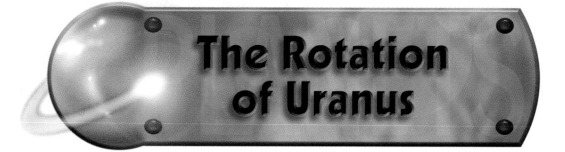

The Rotation of Uranus

Like all the other planets, Uranus travels around the Sun. The nine planets move in paths called **orbits**. It takes Uranus 84 Earth years to travel once around the Sun.

The planets also spin as they travel around the Sun. Each planet spins on an **axis**. An axis is an imaginary pole that runs through the center of a planet. The spinning of a planet on its axis is called **rotation**. It takes Uranus 17 hours and 15 minutes to complete one rotation on its axis.

Uranus is the only planet that spins on its side. Uranus has an axis that points almost **horizontally** in relation to its orbit. Each of the other planets in our solar system has an axis that stands up almost **vertically**. Astronomers do not know if Uranus was made that way or if it was knocked onto its side. Can you imagine anything powerful enough to knock a large planet on its side?

It takes Uranus 84 Earth years to travel around the Sun. Only two planets, Neptune and Pluto, take longer.

The Atmosphere on Planet Uranus

Many different gases form the outer layer of Uranus. **Hydrogen** and **helium** are the most common. Together they make up about 90 percent of the planet's atmosphere, or the layer of gases that surrounds it. Methane also makes up part of Uranus's atmosphere. Underneath this layer of gas is an ocean made up of water, **ammonia**, and liquid methane. Some scientists think that underneath the water is a small, solid **core**. These scientists think that the core of Uranus is made up of heavy, rocky material. There are other scientists who do not believe that Uranus has a core of any kind.

There are thick clouds blowing across Uranus. The clouds are a blue green color because they have methane gas in them. Methane gas looks green in the sunlight.

14

Hydrogen and Helium

Water and Ammonia

Rocky Core

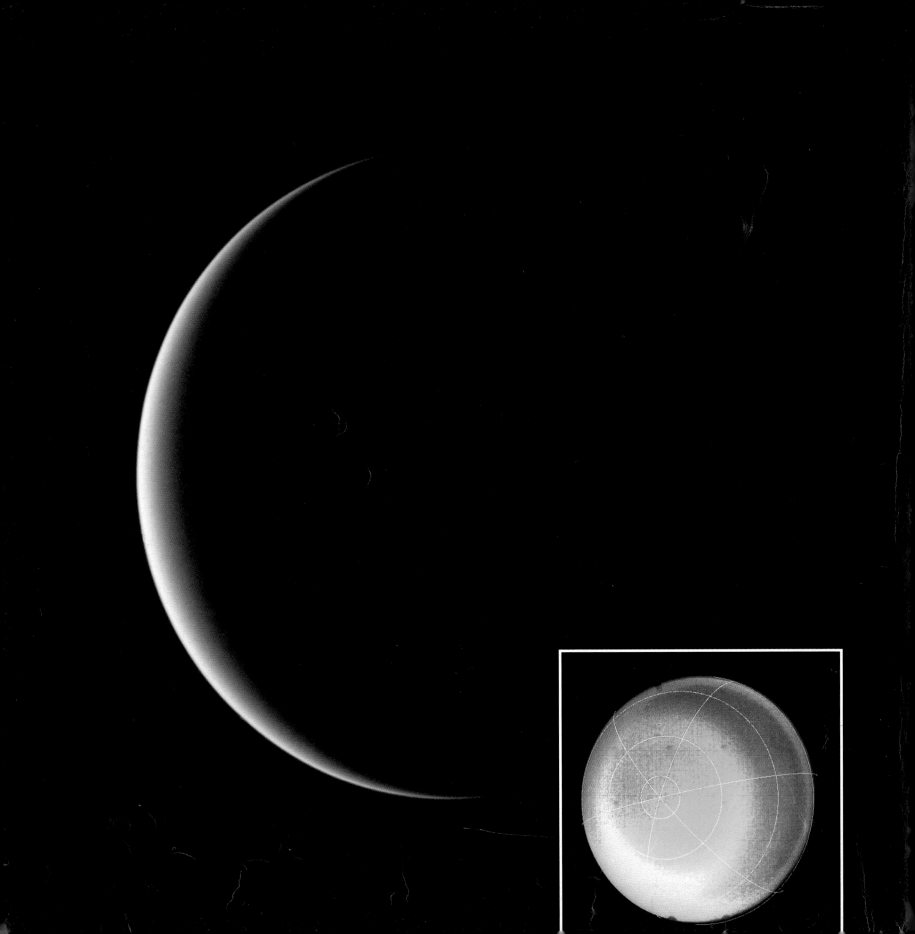

Voyager 2

In the summer of 1977, two **space probes** called *Voyager 1* and *Voyager 2* were sent out into space. There weren't any **astronauts** in the probes. The probes were steered by scientists on the ground. The space probes were sent to get a closer look at the four gas giants. *Voyager 1*'s mission was to Jupiter and Saturn. *Voyager 2* passed by all four planets. It took *Voyager 2* over eight years to reach Uranus. *Voyager 2* got within 51,000 miles (82,076 km) of the planet. That may seem far away, but it is really quite close compared to Uranus's distance from Earth. *Voyager 2* was able to take pictures of Uranus. Using these pictures, scientists were able to figure out what materials Uranus was made of and how these different materials fit together.

These images of Uranus were taken by Voyager 2. The lines on the right image show that Uranus's atmosphere moves in the same direction as the planet.

The Moons of Uranus

Before *Voyager 2's* journey, we already knew about five of Uranus's moons. Those moons were big enough to see using telescopes from Earth. In 1986, *Voyager 2* discovered 10 smaller moons traveling around Uranus. Then in 1997, *Voyager 2* showed that Uranus had two more moons. The **Hubble Space Telescope** was **launched** in 1990 and is still in space. It has sent back many pictures of Uranus and other planets. Pictures of Uranus's two additional moons meant that the planet had a total of 17 moons. Only Saturn, with 18 moons, has more moons than Uranus.

Many of Uranus's moons were named after characters in plays written by William Shakespeare.

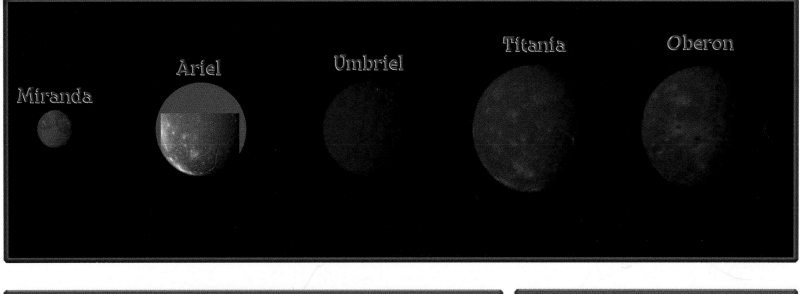

Miranda   Ariel   Umbriel   Titania   Oberon

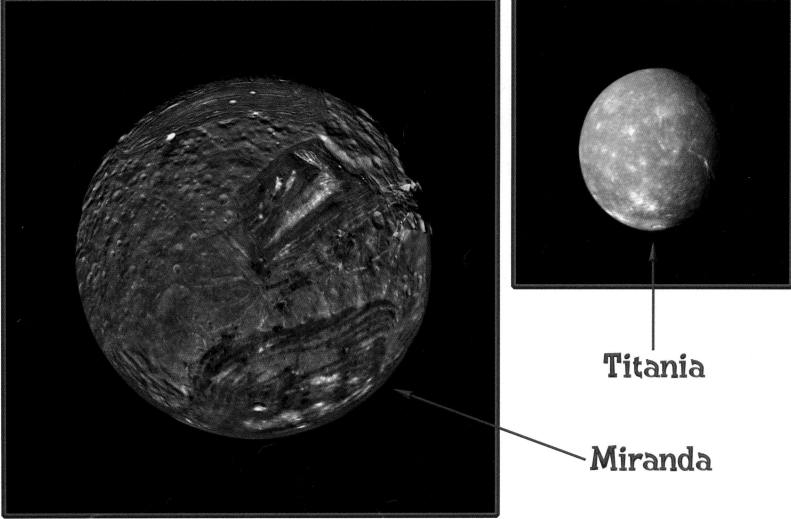

Titania

Miranda

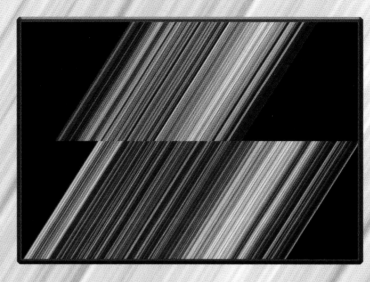

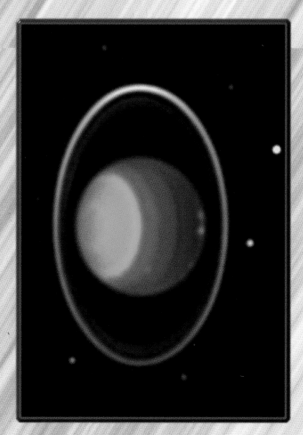

Uranus's Rings

Voyager 2 also took pictures of the **rings** around Uranus. All the giant planets have rings around them. Uranus has at least 10 rings. The ring closest to Uranus is 26,000 miles (41,843 km) from the planet's center. Most of the rings are about six miles (10 km) wide. That is small compared to the size of the planet. The rings around Uranus are different from the rings that circle Saturn and Jupiter. Uranus's rings aren't as wide. This is because they are made of different materials. The rings around Uranus are most likely made of rock, ice, and **carbon**. The rings are almost black in color. Each ring is made of millions of fragments. Fragments are broken up pieces of material. The fragments that make up these rings are sometimes very small. They can also be as big as large boulders.

The image on top is a close-up view of one of Uranus's rings. The bottom picture shows eight of the 10 rings around Uranus.

Planet Uranus's Future

There are no future missions planned to visit the planet Uranus. However, the Hubble Space Telescope keeps sending back pictures of the planet. In March of 1999, the Hubble Telescope sent back pictures of Uranus's different seasons. It showed bright clouds on the planet. It also showed rings that shook like a loose wheel on a wagon.

A mission to the planet Pluto is now being planned. To reach Pluto, a space probe will have to fly through Uranus's orbit. There is no doubt that astronomers will continue to study the icy, blue planet called Uranus.

Glossary

ammonia (uh-MO-nyah) A colorless gas, made up of nitrogen and hydrogen, that has a strong smell.

astronauts (AS-troh-nots) Members of a crew on a spacecraft.

astronomer (uh-STRA-nuh-mer) A scientist who studies the night sky and the planets, moons, stars, and other objects found in space.

axis (AK-sis) A straight line on which an object turns or seems to turn.

carbon (KAR-bon) A chemical found in all living things.

core (KOR) The center of a planet.

gas (GAS) A substance that is not liquid or solid, that has no size or shape of its own, and can expand without limit.

helium (HEE-lee-um) A light, colorless gas.

horizontally (hor-ih-ZON-til-ee) Going from side to side.

Hubble Space Telescope (HUH-bul SPAYS TEL-uh-skohp) A telescope launched into space in April 1990 that has sent back many images of Uranus and other planets.

hydrogen (HY-droh-jen) A colorless gas that burns easily.

launched (LAWNCHD) Pushed out or put into the air.

methane (MEH-thayn) A colorless, odorless gas.

orbits (OR-bits) The circular paths traveled by planets around the Sun.

rings (RINGZ) Thin bands of rock, ice, and other organic material that stretch around the four giant planets.

rotation (roh-TAY-shun) The spinning motion of a planet around its axis.

solar system (SOH-ler SIS-tem) A group of planets that circle a single star. Our solar system has nine planets, which circle the Sun.

space probes (SPAYS PROHBS) Small spacecraft that travel in space and are steered by scientists on the ground.

telescope (TEH-luh-skohp) An instrument used to make distant objects appear closer and larger.

vertically (VER-tih-kul-ee) Going up and down.

Index

Web Sites

If you would like to learn more about Uranus and the other planets, check out these Web sites:
http://www.tcsn.net/afiner
http://msgc.engin.umich.edu